Pool and Spa Leak Investigation

Subtitle: A Practical Guide for Trained Professionals

David Watson

The swimming pool bloke who reads a lot

Publisher: David Watson Pool Consulting

ISBN (Paperback): 978-1-7645296-0-0

ISBN (eBook): 978-1-7645296-1-7

Printed in 2026

Publisher: David Watson Pool Consulting
www.davidwatsonpoolconsulting.com

Cover design by Sue Knox

Illustrations by Sue Knox

Ingram Spark / Amazon KDP Print-on-Demand Edition

In preparing this handbook, AI tools have assisted with editing and verifying references across multiple international databases. All content has been carefully reviewed and edited for accuracy and reliability, informed by my four decades of field experience and current regulatory standards.

Acknowledgement

I extend my sincere thanks to the early readers who reviewed the initial drafts of this book. Your thoughtful feedback, careful reading, and constructive suggestions made this work stronger and clearer. I am grateful for your support in developing this manuscript.

Foreword

I wrote this pocket guide because many leak investigations start with the wrong assumptions and end with the wrong conclusions. A pool can lose water for perfectly normal reasons—bathing uses, wind, evaporation, backwash bypass, or a jammed autofill—and a rushed test can create false positives. At the same time, I've seen texts recommending pressure-testing with compressed air in the field. That practice is unnecessary and dangerous. This guide is my attempt to put a safe, repeatable, and defensible process into the hands of trained pool and spa professionals.

The methods here are deliberately practical. They focus on identifying and confirming water loss using observation, evaporation testing, dye testing, and water-only pressure-testing, with simple temperature correction where needed. I'm not trying to cover every possible system or repair method—those sit outside the scope of a small field guide. My goal is to help technicians communicate clearly with clients, document results in a way that stands up to scrutiny and reduce time wasted chasing 'leaks' that aren't leaks — or worse, creating false ones.

How to Use This Guide

Who This Guide Is For

This guide is written for trained pool and spa professionals, service technicians, supervisors, and maintenance staff responsible for diagnosing water loss in residential, commercial, and public aquatic systems. It assumes basic competency in hydraulics, isolation, safe work practices, and familiarity with pool equipment.

It is not intended for untrained operators, homeowners, or anyone performing work outside the licensing requirements of their region.

How to Use This Guide in the Field

This is a field handbook, not an engineering manual. Each chapter builds a simple, repeatable workflow:

1. **Start by confirming whether a leak is likely**

   False positives waste time. Use the pre-checks and evaporation test to rule out normal water loss before escalating.

2. **Move from the lowest risk → to the highest certainty**

   - Observation
   - Evaporation test
   - Dye testing
   - Water-only pressure-testing

   Each step is designed to eliminate assumptions and avoid unnecessary excavation or repairs.

3. **Use the tests to narrow the problem, not guess**

   The guide shows how to interpret changes between equipment-running and equipment-off tests, how to isolate circuits, and when results point toward shell leaks vs plant leaks vs pipework.

4. **Document everything**

   Use the included record sheets and pressure-test certificate. Clear documentation protects both the technician and the client if the job becomes contentious.

This guide works best when followed in order, without skipping steps, and when the technician slows down enough to let each test stabilise.

Explaining Results to Clients

Clients often expect binary answers: "Is it leaking or not?"

Leak testing can produce three legitimate outcomes:

- **Confirmed leak**

  A test shows a measurable, repeatable loss.

- **Unlikely leak**

  Both evaporation tests match or stabilised water-only pressure-tests show no meaningful pressure drop.

  Explain that pools lose water naturally from wind, splash-out, autofill issues, and evaporation. Show the recorded values, so the client can see the comparison themselves.

- **Inconclusive**

  Air entrapment, poor isolation, weather, or site conditions can prevent a definitive result.

  Explain that inconclusive does not mean no leak — it simply means the test needs to be repeated under better conditions, or isolation needs to be improved before proceeding.

A few months ago, we set up an evaporation test on an indoor pool. It happened that there was torrential rain overnight whilst the evaporation test was running. The result of the evaporation test was inconclusive, as the louvre windows were missing a seal; the torrential rain had come in and run down inside the windows, causing the pool and gutter to fill with rainwater. We had to repeat the test in better weather. Don't assume that, as a pool is indoors, the weather can't affect it.

Your job isn't to guess. Your job is to present the results clearly, explain what they mean, and outline the next logical step. This builds trust, manages expectations, and avoids the industry's habitual leap from suspicion to excavation without evidence.

Table of Contents

1. Introduction

> This guide is intended for trained pool and spa professionals. It does not replace site-specific risk assessments or local regulatory requirements. It provides practical, safe, and repeatable field methods to help technicians identify and confirm water loss in pools and spas.

This guide focuses on field methods for identifying and confirming water loss. It does not cover:

- structural crack assessment, shell movement analysis, ground movement engineering, hydrostatic uplift calculations, or long-term repair design
- repair methods (pipe repairs, skimmers, light niches/conduits, sealants, epoxies, grouts, linings, resurfacing)
- specialist detection tools (acoustic devices, correlators, tracer gases, infrared, CCTV)
- licensing, legal authority, insurance documentation, or contractual interpretation

Where conditions fall outside the scope of this guide, consult appropriately qualified professionals

Why Water-Only Pressure-Testing

It's surprisingly common to see pressure-tests done using air or other gases. This is a problem for two reasons.

First, air (and any gas) will compress. In a water-based test, if you set the pressure to 200 kPa and there's air trapped in the system, that air continues to compress. As it compresses, the gauge shows the pressure dropping. That drop can look like a leak and falsely invalidate an otherwise sound test.

Second—and far more important—compressed air stores energy in a very different way than water. When it lets go, it can let go all at once. The result is often explosive. A sudden failure in a water-only test usually means a slump of water and a noise as the joint fails. The same failure with compressed air can send pieces of plastic flying in any direction, including straight at you.

Years ago, when I was working in quality control in the plastics industry, we used air for product testing. The test item was bolted into a rig so it couldn't move. The airline was connected. The gates to a special enclosure were shut. We sat two metres back from the enclosure, wearing face shields. That level of protection was required for a reason: air-pressure failures can be catastrophic.

The methods contained in this guide provide a safe, repeatable, and defensible water loss test method without the risks associated with compressed gases. The test method and temperature correction outlined in this document align with:

- AS 1926.3
- AS 2610
- WHS regulations
- Plumbing safety principles

Local regulatory requirements may differ; users are responsible for confirming applicability in their jurisdiction.

At the time of writing in 2026.

> Never use compressed air or gas for pressure-testing in the field.

What This Guide Won't Solve (Read This First)

This guide is a field method for confirming water loss and pressure-testing pool and spa pipework using water-only. The method is solid when it's applied properly.

What this guide won't do is save you from bad conditions, bad isolation, or bad judgment. Paper can't see your site. You can.

Use this page as your "stop and think" moment.

It won't fix poor isolation or hidden interconnections

This guide assumes the line you're testing is isolated.

It won't protect you if the system is still connected to things like:

- booster/jacking pumps

- shared plant rooms

- balance tanks and overflows that quietly feed back

- auto-fill or make-up water that isn't properly shut off

- valves that don't seal, or pipework that has been modified over the years

If there's any path for pressure or water to enter from outside your test setup, your results can be wrong, and the test can become unsafe.

It won't prevent false "fails" if air is trapped

Water doesn't compress. Air does.
If air is left in the line, the gauge can drop and make a sound line look like it's leaking.

A slow, drifting pressure loss—especially after you first pressurise—is often air working its way out, not a leak.

If you can't stabilise the pressure after bleeding and re-bleeding, treat the result as inconclusive, not failed.

It won't stop things from popping out if you set up carelessly

Even with water-only testing, a bung or fitting can come loose if it's poorly seated, undersized, or not restrained.

This guide won't prevent injury if you:

- stand over the bleed point

- stand in the line of fire of a bung

- rush the setup

- assume "it'll be right"

Water-only testing reduces stored energy compared to air, but it does not eliminate risk. Wear eye protection, stand to the side when bleeding, and treat every pressurised fitting with respect.

It won't replace engineering or specialist assessment

This is a leak investigation field guide. It does not diagnose:

- structural cracking or shell movement

- ground movement/settlement

- hydrostatic uplift issues

- long-term repair design or "root cause" engineering

If the job is heading into structural territory, get the right people involved.

It won't solve disputes, insurance arguments, or contract issues on its own

The forms and certificates help with documentation, but this guide is not legal advice, and it won't magically make a disputed job simple.

If the consequences of being wrong are high (insurance, litigation, major defects, commercial handover), slow down and document everything properly—or escalate.

It won't make judgment calls for you

This guide can't tell you when something "doesn't feel right".

If the test behaviour is odd, if site conditions change, or if you don't trust the setup—**stop**. Recheck isolation, rebleed, and reassess.

If you wouldn't bet your own reputation on the result, don't sell it as certainty to the client.

Stop and Escalate If Any of These Apply

- You can't confirm full isolation of the line/system

- Pressure rises without you adding water (external pressurisation is happening)

- Pressure won't stabilise after proper bleeding/re-bleeding

- A bung or fitting won't restrain confidently

- The site is pushing you to rush or skip steps

- The outcome affects insurance, certification, or commercial sign-off

This guide is a tool, not a substitute for competence. Used properly, it gives reliable answers. Used carelessly, it can mislead.

2. Safety and Risk Disclaimer

All procedures described in this guide involve risks associated with water systems, hydraulics, confined spaces, and pressurised equipment. These methods must only be performed by trained professionals familiar with Australian or local WHS obligations and local regulatory requirements.

These procedures assume competency in basic hydraulics, safe isolation, and the use of manual handling and pressurised-system tools.

Technicians must:

- Conduct a site-specific risk assessment before starting any test.
- Wear appropriate PPE at all times, including eye protection, gloves, and enclosed footwear.
- Ensure the work area is clear of bystanders, pets, and clients during testing.
- Keep hands, face, and body away from all bungs, plugs, valves, gauges, and fittings when the system is pressurised.
- Stand to the side when opening any bleed valve — never over the discharge path.
- Use water-only pressure-testing; do **not** use compressed air or gas under any circumstances.
- Depressurise the system fully before adjusting bungs or altering the setup.

Failure to follow safe procedures can result in serious injury, equipment damage, or unintended release of water.

3. Is There a Leak?

Often, a pool or spa is reported to be leaking in error. Before starting leak testing, consider:

Chemical logs, if available	Cyanuric acid, TDS, salt and calcium hardness are only altered by water replacement. If any of these are reasonably stable compared to other pools on your route, chances are there is no leak. Assumes no partial drain, backwash dumping, or chemical dilution events.
Autofill line	If fitted, check that the autofill is not constantly running or jammed and is operating correctly.
Backwash line	Always check the backwash line for leaks before commencing any other leak tests.

Observation	**Wind**
	If the pool is in a windy location or if the surroundings are wet and the weather is blustery, there is a good chance the water loss is due to wind. This weather would make testing difficult.
	Equipment Pad
	Surrounding ground, slab or soils for filter bases or other equipment may indicate leaks from the plant. Attend to these items first before commencing any leak testing.
	Rust Stains
	Rust stains, particularly around the equipment pad or along the edges of the water body, can indicate potential water loss.
	Surrounds
	Surrounds for the pool itself. Depending on the local soil, leaks can show up as puddling or surface dampness. Paving may dip or sink, or other surface imperfections may suggest an area of concern. Grass may be very green or very dead in spots where the leaks may be occurring.

Weather	Sudden weather changes can cause a water body to evaporate more. Not only hot weather, but also sudden wind can cause water loss.

There may be other factors to consider in your local area; this is a generic list to start.

> Note: Sometimes water needs to be checked to work out if the leak being seen is pool water or not. Say a continuous dripping outside the pool in the stairwell. For an outdoor pool, the check to use is cyanuric acid. Cyanuric acid does not get added to any other water source, will not be added by travelling through concrete or other sources and won't react to organics like chlorine would. Collect the water using a container and complete a cyanuric acid test; if present, it is pool water, if not, it isn't.

4. When Not to Test

Delay testing if:

- High winds (makes testing unreliable and unsafe)
- Heavy evaporation conditions
- Autofill not isolated
- Equipment leaks unresolved

5. Equipment

Pressure Cross

Irrigation manufacturers produce a polyethylene cross, usually in 20 mm or 25 mm. This fitting is extremely helpful for pressure-testing pipes.

A cross gives four connection points:

1. Connection to the pipeline being tested

2. Connection for the water supply (typically a garden hose)

3. Air-release valve to bleed off entrapped air

4. Pressure gauge connection

Because it's a cross, it can be rotated as needed and still allow the air-release valve to sit at the highest point. Being a 25 mm fitting, it can be adapted to almost any pipe size, and garden hoses (usually 15 mm) won't be flow-restricted.

A liquid-filled pressure gauge is usually more stable and easier to read than one that isn't. If purchasing a gauge specifically for leak testing, opt for a liquid-filled gauge.

Note: Interesting aside. Liquid-filled gauges use glycerol as the liquid. This can be changed or topped up from time to time.

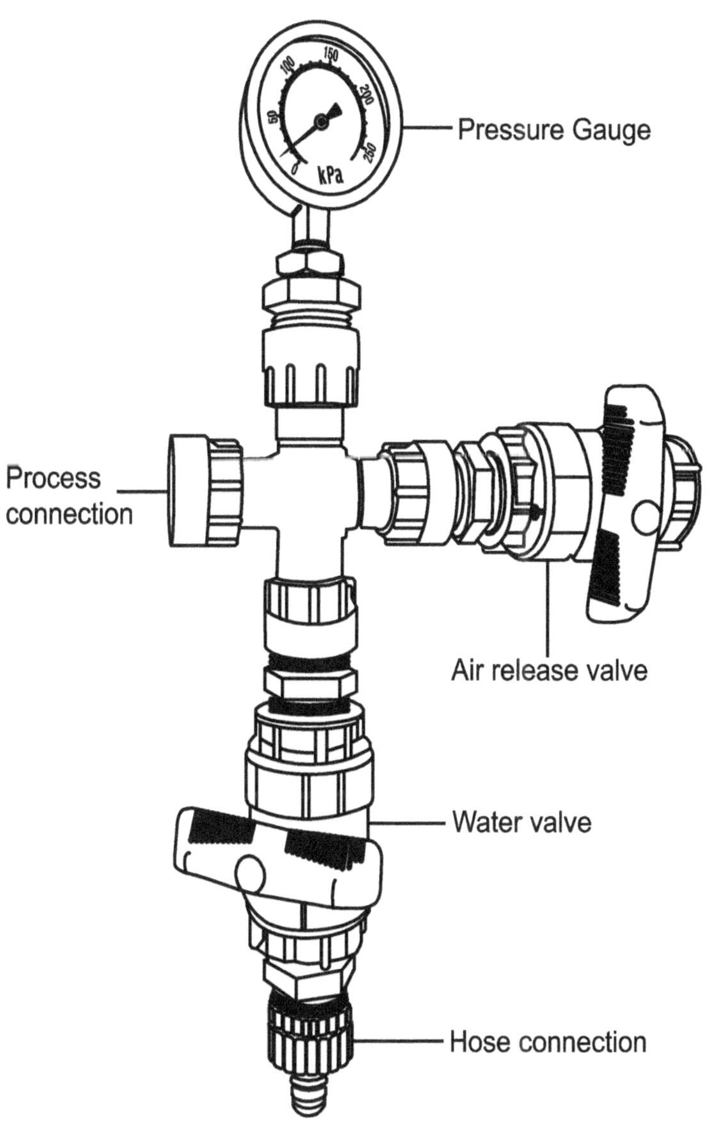

kPa
Pressure Gauge
Process connection
Air release valve
Water valve
Hose connection

Bungs

Bungs are used to block off pipe openings wherever they occur. There are different types of bungs available, and the job will determine which is best suited.

Typical tapered swimming pool bung. They can be found without tapering as well.

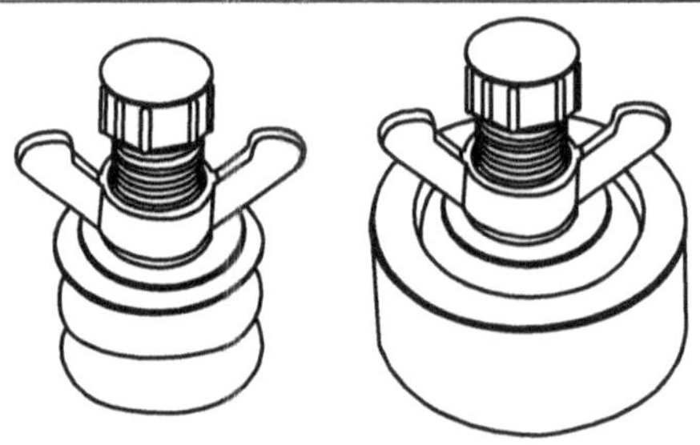

Sometimes referred to as test plugs, this style of bung is primarily sold in plumbing stores. The main advantage is that the threaded section can be used to mount the pressure cross or valves to bleed air from the pipework.

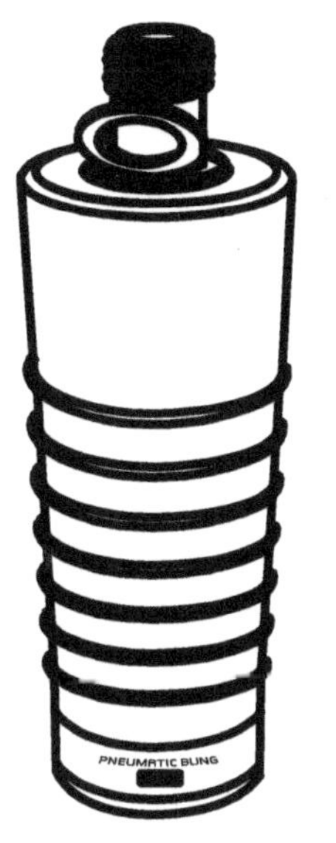

This pneumatic type of bung is sometimes preferable. The main advantage is that they are flexible and can be pushed into pipes that may not easily take any of the other bungs.

Note: For larger pipework 200 mm plus, this type of bung is usually easier to work with.

Basin

For evaporation testing, I prefer a 2-litre plastic ice-cream container. They're lightweight, easy to carry, common, and usually discarded once empty. The thin plastic transfers temperature quickly, which helps equalise temperatures.

To keep the basin from drifting, I punch a small hole in a top corner and run a cable tie through it to make a loop. A short piece of string ties it to something on the pool deck.

6. Evaporation Testing

Backwash Check

Before running an evaporation test, run the circulation system and check the backwash line for leaks.

To check properly, loosen the backwash union enough that you're sure there's no water trapped under pressure. Let it drain fully. Restart the pump and watch for a steady stream. If you see flow before starting an evaporation test, fix that leak first.

Evaporation Test

If the pool is outdoors, check the weather. You need no rain, no showers, and ideally no wind.

Do not perform this test during heavy bather load periods. Ideally close the pool to bathers for the test period.

Discuss with occupants that the pool equipment needs to run overnight.

Steps:

1. Fill the pool to its normal operating level.
2. Isolate any autofill device and check it is not running.
3. Turn circulation on manually.

4. Turn the salt chlorinator off or reduce output.

5. Mark the pool water level with a pencil.

6. Half-fill to three-quarters-fill the basin and mark the level.

7. Float the basin in the pool, so its temperature matches the pool. (If the basin is touching the pool wall, convection heat transfer may skew results)

8. Tie it off so it doesn't drift.

9. Leave for at least 12 hours (I prefer 24). Measure the drop in both pool and basin:

   a) Same drop: no leak.

   b) Pool drops more: continue testing.

   c) Basin drops more: basin was compromised; repeat it.

After testing with equipment-running, repeat with equipment-off for further information.

Test	Outcome	Likely Conclusion
Evaporation test plant running	No difference in measurement	Not likely leaking
Evaporation test plant running	Difference in measurement	Check the plant thoroughly for potential leaking.

Test	Outcome	Likely Conclusion
Evaporation test plant not running	No difference in measurement	The pool shell and pipework are unlikely leaking.
Evaporation test plant not running	Difference in measurement	Consider letting water drop to the lowest level
Evaporation test, both running and not running	No difference in measurement	Not likely leaking
Evaporation test, both running and not running	Difference in measurement	Consider letting water drop to the lowest level and pressure-testing the pipework.

7. Low Level Test

The objective of the low-level test is to find the leak by simple means. Low-level testing is an *indication* tool, not a confirmation method.

The filled water body is allowed to leak to the level it stabilises at. The assumption is that a point just above this is the point where the leak is most likely to be. A common leak point found using this method is the pool light.

This method can be deceptive. A leak from an expansion joint at the bottom of the water body will rapidly drain when the pool is full; however, as the water level above it decreases, it will lose less and less water, and this can make the level look stable even when it isn't. The reduced water loss is a direct result of the pressure above the leak point. For example, if the pool is full at 2 metres deep, this is 20 kPa; if the water level falls to say 1 metre deep, the pressure is halved to 10 kPa, and the leak will slow substantially.

An alternative to the low-level test is a dye test if a light or fitting is suspect. The dye test can also be used on an expansion joint; however normally requires scuba to be successful.

8. Dye Test

As the name implies, this test uses a dye to track the water movement and hopefully find the leak point.

This is useful where lights or fittings are suspected as the likely cause.

This test is best performed in no-wind conditions, as wind will make it hard to see the results.

It is imperative to be sure that no equipment is running. Even a slight water current caused by a running heater, negative-edge flow, or solar return can invalidate dye tracking.

1. Turn off all equipment on the water body.
2. Fill the water body as high as possible. We want a lot of water above the leak for this test.
3. Using food-grade dye or phenol red pH indicator with a chlorine neutraliser, via a syringe, very gently, without disturbing the water, squirt a small amount under the water near the suspect leak point.

> Note: Phenol red is commonly used because technicians
> carry it for pH testing. Use sparingly. Food-grade dye
> (available at supermarkets) is an equally effective
> alternative. Before the phenol red dissipates into the
> water, it may show signs of tracking.

4. Follow the track until it is certain that the dye is tracking into the light or fitting surrounds.

5. Do not stop at the first confirmed leak. Continue to check all the fittings and lights in the entire water body that are reachable.

6. It is tempting to enter the water body to conduct this test. It can be done this way; however, remember that your movements in the water can impact the test.

7. If you are scuba qualified, and it is safe to do so, as in the depth is easy to stand up in and be clear of the water, then the dye test can be used around any fittings or joints in the bottom of the pool. Never conduct scuba dye testing alone. This cannot be successfully conducted using duck diving techniques, as this disturbs the water too much for accurate testing.

9. Pipe Gravity Test

This is a less common test. It is not as effective as the other tests and is normally only employed when no other test can be successfully conducted. Examples of this could include pipe work from the gutter to a balance tank.

To conduct this test:

1. Bung off the accessible end of the pipe if possible and safe to do so.
2. Fill the pipe with water to a point that can be observed and seen.
3. Mark the water level in the pipe where it can be seen.
4. Wait 24 hours and measure the new water level from the original mark. Hopefully, it is a very minor difference.

This test will work best if there are two or three pipes tested at the same time. It is unlikely that all the pipes will drop by an equal amount if there is a leak. A basin could be considered to act as a witness; however, be aware of the potential temperature difference encouraging the basin to evaporate faster.

Because temperature and wetting effects can shift the visible water line, treat this as an indication only.

Note: This is not as effective as the other methods and is not a robust method. It should only be considered when no other option exists.

10. Safe Pool Industry Pressure-Testing

Pressure-testing in swimming pool and spa work should be done using water. Using air or other compressed gases is not recommended, as it can introduce unnecessary risk. Compressed gases can store significant energy. If a fitting, plug, or test cap were to loosen or fail, that stored energy can be released suddenly. Because water is effectively incompressible, any leak or failure typically results in a less forceful and more manageable release.

A key practical step in achieving a reliable test is to remove as much trapped air from the line as reasonably possible. There are a few accepted methods for doing this, but a straightforward approach is to install bleed points (bleed plugs/valves) at the highest points in the pipework so that air can be purged during filling.

Where the system has a single clear high point, it is often most effective to fill and pressure-test from that location, as it supports better air removal and more consistent test results.

To conduct a pressure-test, a pressure-test cross is best set up before commencing. See chapter 4 on how to set up the cross.

Note: Whilst not mandatory by any means, a calibrated pressure gauge is worth considering for this application. Be aware that if under contract, this may be required; check contract wording.

Once the pressure cross is assembled and checked for water tightness, it is possible to commence setting up for the pressure-test. The following are suggested steps to follow (note the steps may need modification on site):

1. Set the pressure cross up on the highest point of the pipeline. Ensure the air release valve is the top valve in the stack.
2. Attach the garden hose or other water supply and start it at a trickle pace.
3. As water is noted escaping from a pipe end, push the appropriate rubber bung into place to block the pipe. If the bung has a test point, leave it open to help remove air more quickly, closing the test point only when water is escaping.

For some pipe setups, it may be better to install the bung before filling with water. This can only be determined on site.

4. Continue with 3 as required to block off all pipe ends and let the water fill the pipeline.

5. When there is a continuous flow of water out of the air release valve on the pressure cross close the valve and monitor the pressure gauge.

> A stable test requires removing entrapped air — the most common cause of false failures.

6. When the pressure gauge reaches the target pressure, for example, 150 kPa, close the valve feeding water into the line and then turn off the water source.

7. Record the time of day, the pressure and the ambient temperature.

8. Monitor the pressure gauge for 30 minutes and ensure there is no movement from the target pressure recorded. If the pressure drops slowly, then it is likely that air is still entrapped in the system. If the pressure drops rapidly and bottoms out, it is most likely a leak.

9. If air entrapment is suspected, restart the water and open the water valve on the pressure cross.

10. Open the air release valve. Stand to the side of the valve when opening it and never directly above the discharge path. Note that water will shoot out of this valve with considerable pressure; eye protection is mandatory. Even modest pressure can eject water several metres.

11. Hopefully, when the column of water settles, it will be bubbling water from the air release valve. Again, close this valve and increase the pressure as at step 6.

12. Once the pressure is holding steady and air entrapment is minimised, the pressure-test can be left for a period. This period will depend on the intended reason for the pressure-test. The following are for guidance only:

   a) Leak detection – 20–30 minutes
   b) Verifying new work – 60 – 90 minutes
   c) Commercial new work – overnight

13. Once the pressure-testing is complete, record the date and time, the end pressure and the ambient temperature.

14. If the project requires a pressure-test certificate, then correction for the temperature is likely required, particularly for overnight results. Please refer to the process for calculating the endpoint for a pressure-test certificate.

> Note: Whilst not required, it is my practice to leave the pipework at pressure on new builds. The idea is that if the pipe is damaged—such as by a star picket during construction—the pressurised water will spray, alerting nearby workers. It is also something that can be checked on returning to the site with moderate pressure drops over time, likely acceptable, but drops to zero require investigation.

11. Temperature Correction for Pressure-Testing

Pressure is impacted by temperature. The pressure will change depending on the temperature, and this can lead to some concern when viewed without correction. Pipe size and length do not materially affect the temperature-corrected pressure result, because the small compressible volume—not the water volume—dominates the pressure change.

> For most pressure-tests, being shorter duration tests, unless there is a significant change in temperature, correction for temperature is not likely required. If conducting a test overnight, always run through the temperature correction.

To correct a pressure, we need to work with the ideal gas law. Below is an example worked through in full on how to perform this calculation. For this example:

- Initial pressure 150 kPa
- Initial temperature 22°C
- Morning pressure: 147 kPa
- Morning temperature: 18°C

The first step is to convert the temperatures to Kelvin (K). Kelvin is −273.15°C In scientific terms, absolute zero.

To convert any temperature to K, all we do is add 273.15°C to our ambient temperature in Celsius. Using our example:

Initial temperature 22°C becomes:

$$22°C + 273.15°C = 295.15\ K$$

Morning temperature of 18°C becomes:

$$18°C + 273.15°C = 291.15\ K$$

Now our data is:
- Initial pressure 150 kPa
- Initial temperature 295.15 K
- Morning pressure: 147 kPa
- Morning temperature: 291.15 K

The formula to apply is:

$$\frac{P_1}{T_1} = \frac{P_2}{T_2}$$

This isn't as hard as it looks. Our data becomes:

- P_1 Initial pressure 150 kPa
- T_1 Initial temperature 295.15 K
- P_2 Morning pressure: 147 kPa
- T_2 Morning temperature: 291.15 K

This can be rearranged to solve for the corrected pressure:

$$P_{2\ corrected} = P_2\ x\ \frac{T_1}{T_2}$$

So now we can run through with our example how to solve:

$$P_{2\,corrected} = 147\ kPa\ (P_2)\ x\ \frac{295.15\ K\ (T_1)}{291.15\ K\ (T_2)}$$

$$P_{2\,corrected} = 147\ kPa\ (P_2)\ x\ 1.013738622703074$$

$$P_{2\,corrected} = 149.0196 kPa$$

So, our pressure drop overnight was:

$$150\ kPa - 149.0196\ kPa = 0.9804\ kPa$$

This minor difference could be caused by pressure gauge inaccuracy, minor air entrapment in the pipeline and similar.

If issuing a pressure certificate, this would be the best way to lay the data out. For a practical rule of thumb for pool work, a pressure-test band of about 2-3 % is likely an acceptable pressure drop. If contract specifications or local codes specify tighter tolerances, always follow those requirements. Anything larger is worth investigation; anything less can be considered no significant leak detected.

Pressure test kPa	Typical Acceptable Drop (kPa)
100	2 -3
150	3-5
200	4-6
300	6-9

12. Practical Examples

Applying the Tests

This is not intended as a definitive list of every situation. It is intended as a guide, and the ideas presented can be applied to most situations. It is assumed that all the pre-checks have been done to determine whether a leak test is worthwhile.

Standard Pool

A standard pool is defined for the purpose as being a typical backyard pool with a skimmer box and 2 or 3 returns, with perhaps a pool light.

For this, pool testing is fairly straightforward:

- Equipment-running evaporation testing
- Not running evaporation testing
- Pressure testing of each line as required

After conducting all these tests, assuming there is a leak, it will be isolated to a specific area or pipe.

Interpreting the results:

- Water loss during equipment operation (running): focus on plant, waste line, multiport, or pressure-side pipework.

- Water loss during equipment shutdown (gravity or no flow conditions): focus on shell penetrations such as lights, skimmer throat, returns, or expansion joints. Pressure-test pipework.

- Failed pressure-test on a single line: check all accessible fittings on that circuit before assuming buried pipe failure.

Overflow Spa and Pool

The overflow spa can be more difficult to test, and in some cases, isolation can be difficult.

Again, though, as in the standard pool, the order of testing doesn't change. However, it may be:

- The spa section needs testing separately from the pool. This may involve additional bungs to isolate the spa. Even for the evaporation tests.
- Likewise, the pool may need testing separately from the spa. Similarly, any interconnecting pipe works will need to be bunged

Interpreting the results:

- Loss only when spa is isolated: focus on spa shell, fittings, or interconnecting pipework.

- Loss only when pool is isolated: treat as a standard pool leak.

- Loss in both isolated tests: consider shared pipework or common plant components.

Infinity Edge Pools

Infinity edge pools are pools with a gutter or spillover that captures the water from the pool. Whilst the order is the same as the standard pool, it is usually necessary to conduct the tests separately.

Test the pool itself and separately test the gutter or spillover catchment when the equipment is not running. During equipment-running tests, this is a normal pool. However, the water level is monitored in the spillover catchment or gutter.

Water levels between weir flow and gutter capture can change with wind. Observe visually before assuming leakage.

Interpreting the results:

- Loss during equipment-off testing of the gutter system suggests the gutter shell, gutter fittings, or balance pipework.

- No loss when equipment is off, but loss when running suggests circulation or waste-related losses rather than shell leaks.

Note: Another leak can be rebound. This is where the gutter or spillover is not capturing the water completely, or the water is hitting something and bouncing away over the edge. Typically, clearing away any ground covering will reveal significant saturation if this is the cause.

Pools With Balance Tanks

Pools with balance tanks are very similar to infinity-edge pools. There are only a few differences.

Many years ago, some pools had an equaliser pipe at the bottom of the balance tank between the pool and the balance tank. Some of these pools are still in existence and are occasionally seen. Originally, there was a float with a flap arrangement as a sort of one-way valve. If the float dropped, the valve opened, and water flooded from the pool into the tank. If the tank has this feature, it is likely easier to bung this from the pool end.

Conducting a test in a pool with a balance tank can be a bit daunting on the first attempt.

The main points are:

- During the equipment-running evaporation test, the basin needs to be floating in the pool or in the balance tank. However, the water level is measured in the balance tank, not in the pool.
- During the equipment NOT running evaporation test, the basin needs to be floating in the pool, and a second basin needs to float in the balance tank.

This is because the leak could be from the balance tank itself, not just the pool.

Interpreting the results:

- Loss observed in the balance tank only: investigate tank shell, penetrations, or equaliser pipework.

- Loss observed in the pool only: treat as a poolside leak.

- Loss observed in both: shared hydraulics or interconnecting pipework should be suspected.

13. Glossary

Autofill

A device that automatically adds water to a pool or spa to maintain a set water level.

Backwash Line

A pipe or outlet used to expel dirty water from a pool filter during cleaning cycles.

Balance Tank

Sometimes referred to as a surge tank. A tank that captures bather displacement, contains water for backwashing and acts as a filling point on commercial pools.

Bung

A plug used to seal the end of a pipe or fitting during pressure-testing.

Chlorinator

A device that automatically dispenses chlorine or sanitising chemicals into pool water.

Dye Test

A method using coloured dye to visually trace water flow and locate leaks.

Evaporation Test

A procedure to differentiate between water loss due to evaporation and actual leaks.

Expansion Joint

A flexible joint in pool construction that accommodates movement and prevents cracks.

Hydrostatic Uplift

The upward pressure exerted by groundwater on the pool structure.

Multiport Valve

A valve that directs water flow through different filter or waste paths in a pool system.

Phenol Red

A pH indicator dye used in water testing, sometimes applied in leak detection.

Pressure Cross

A polyethylene or PVC fitting with four ports is used to connect the supply, gauge, air release valve, and the pipeline for pressure-testing.

Pressure Gauge

An instrument used to measure the pressure within pool pipes during testing.

Scuba Dye Testing

Dye testing is performed underwater by a certified scuba diver to locate leaks in inaccessible areas.

Sight Glass

A transparent section in a pipe or valve used to observe water flow.

Temperature Correction

Adjusting pressure readings to account for changes in temperature using the ideal gas law.

Trapped Air

Air pockets within pool pipes that can affect pressure-test results.

Water-Only Pressure-Testing

A safe pressure test method using water rather than compressed air or gases.

14. Imperial Appendix

This appendix provides imperial unit equivalents and conversions for use where imperial-based documentation or equipment is encountered. The primary guidance in this book is based on SI units.

Units – kPa to PSI

Kilopascals (kPa)	Pounds per Square Inch (PSI)
100	14.5
150	21.8
200	29.0
300	43.5

Rule of thumb: 1 PSI ≈ 6.9 kPa, 1 kPa ≈ 0.145 PSI.

Temperature Conversion – Fahrenheit to Kelvin

To convert Fahrenheit (°F) to Kelvin (K):

K = (°F − 32) x 5 / 9 + 273.15

Example:

68°F → (68 − 32) x 5 / 9 + 273.15 = 293.15 K

Imperial Pressure Test Certificates

In regions using imperial units, pressure-test certificates may record pressure in PSI and temperature in °F. When interpreting or converting results:

- Convert pressures to kPa before applying acceptance bands used in this guide.
- Convert temperatures to Kelvin before applying temperature correction calculations.
- Clearly document both the original imperial values and the converted SI values on the certificate.

15. Evaporation Test Record Sheet

Purpose: distinguish evaporation from leakage by comparing pool water loss to basin/bucket loss.

A fillable PDF version is available at:
www.davidwatsonpoolconsulting.com

SITE / CLIENT	
ADDRESS	
TECHNICIAN	
DATE(S)	

POOL TYPE			
☐SKIMMER	☐OVERFLOW	☐INFINITY	☐BALANCE TANK
AUTOFILL ISOLATED?		☐YES ☐NO	

<table>
<tr><td colspan="4">WEATHER NOTES</td></tr>
<tr><td>☐WIND</td><td>☐TEMP</td><td>☐RAIN</td><td>☐FORECAST</td></tr>
<tr><td></td><td></td><td></td><td></td></tr>
</table>

Test 1 – Equipment - Running

MEASUREMENT	START	FINISH	DROP
POOL WATER LEVEL			
BASIN/BUCKET LEVEL			

Notes (equipment status, chlorinator output, any disturbances):

Test 2 – Equipment-Off (If performed)

MEASUREMENT	START	FINISH	DROP
POOL WATER LEVEL			
BASIN/BUCKET LEVEL			

Interpretation (circle or tick):

☐Same drop (pool ≈ basin): evaporation dominant; leak unlikely.

☐Pool drops more than basin: continue investigation; leak possible.

☐Basin drops more than pool: basin compromised; repeat test.

Additional observations:

Sign-off

Technician Name/Signature	
Client Acknowledgement	

16. Water-Only Pressure-Test Certificate (Generic)

Use for documenting water-only pressure-testing of pool/spa pipework. This form does not authorise work outside licensing/regulatory requirements.

A fillable PDF version is available at:

www.davidwatsonpoolconsulting.com

SITE / CLIENT	
ADDRESS	
TECHNICIAN / COMPANY	
CONTACT	
DATE	
POOL/SPA TYPE	

<table>
<tr><td colspan="6">LINE TESTED</td></tr>
<tr><td>☐SKIMMER</td><td>☐MAIN DRAIN</td><td>☐RETURNS</td><td>☐CLEANER</td><td>☐SPA</td><td>☐OTHER
______</td></tr>
<tr><td colspan="3">TEST POINT LOCATION (HIGHEST POINT USED?)</td><td colspan="3"></td></tr>
<tr><td colspan="3">GAUGE TYPE / RANGE / CALIBRATION DATE (IF KNOWN)</td><td colspan="3"></td></tr>
<tr><td colspan="3">WATER SOURCE (MAINS/TANK)</td><td colspan="3"></td></tr>
</table>

Test Setup Confirmation

☐Water-only test (no air or gas used).

☐All accessible ends bunged/isolated and visually checked.

☐Air bled from the highest point until continuous water observed at air-release.

☐Eye protection used; technician stood clear of discharge when bleeding.

☐Auto-fill isolated and plant/waste configuration noted (as applicable).

Recorded Readings

Reading	Time	Pressure (kPa)	Ambient Temp (°C)
Start			
30 min check			
End			

Overnight / Temperature Correction (if used)

Input	Value
T1 (start temp, K)	
T2 (end temp, K)	
P2 corrected (kPa) = P2 x T1 / T2	
Corrected pressure drop (kPa) = P1 – P2 corrected	

Conclusion

☐*Pass*: no significant pressure loss detected after stabilisation and (if applicable) temperature correction.

☐*Inconclusive:* likely trapped air / gauge issue; re-bleed and retest.

☐*Fail:* significant pressure loss suggests leak on tested line; proceed with isolation/dye/section testing as appropriate.

Additional observations:

Sign-off

Technician Name/Signature	
Client Acknowledgement	

17. About the Author

David Watson entered the aquatic industry by accident —
literally. Early for a job interview at a carpet factory in Auckland,
he parked outside a closed pool shop to read a book. An ex-
business partner had barricaded the shop, and when the owner
returned from the Takapuna police station, he assumed David
was involved in the break-in. After a six-and-a-half-hour
conversation, David walked out with a job offer that would shape
the next forty years of his life.

His early training came from an eclectic mix of mentors: private
tuition in water chemistry from Ty Webb, guidance from a
Henderson Environmental Health Officer, and hands-on
instruction from Ken "the poison dwarf" of EziClor Chemicals,
whose own training traced back to the Blue Haven School in Los
Angeles.

After relocating to Western Australia, David spent three decades
with Shenton Aquatics, becoming one of the state's most
experienced specialists in recirculating aquatic systems. His work
spans microbial risk mitigation, hydraulic troubleshooting,
compliance documentation, and technical innovation, including
co-inventing a US-filed water chemistry dosing process.

Across those decades, David has worked alongside practical,
capable pool technicians — the kind who can rebuild a pump with

a shifter and a bit of swearing, ride motorbikes on the weekend, and drink him under the table without trying. David is the opposite: bookish, early-to-bed, and happiest when solving a technical puzzle. This guide is his way of giving something back to the people who taught him the trade from the ground up — a simple, safe, and reliable leak-testing method that doesn't require an engineering degree to use, just a willingness to do the job properly.

www.davidwatsonpoolconsulting.com